Mommy Do Dinosaurs Have Belly Buttons?

Authored by
Laura J. Steele, MA, CHES

Illustrated by
Alissa Mendenhall

ISBN: 1500621994
ISBN 13: 9781500621995
Library of Congress Control Number: 2014913200
CreateSpace North Charleston, South Carolina

This book belongs to

_____

Alexander's mother put him in the backseat of their car. He was very sleepy and rubbed his eyes.

Alexander asked, "Mommy, why did we have to get up so early this morning? It's Saturday, and I wanted to stay in bed and sleep."

She got into the front seat of the car and started to drive. They were leaving their house in Colorado Springs, Colorado.

"Remember, I told you last night that we are going to a dinosaur museum today? We're going to Dinosaur Ridge in Morrison. They have a new exhibit on dinosaur eggs." His mother said, while she smiled and looked at him in the rearview mirror.

"How long do I have to be in the car?" Alexander asked.

"For an hour and fifteen minutes," his mother answered.

Six-year-old Alexander let out a big sigh and was quiet for a while as he played with the toy Tyrannosaurus rex that he held in his hands. Soon he fell asleep.

OLDEST DINOSAUR
BABIES RECENTLY
DISCOVERED

The drive went smoothly, and his mother woke him up by saying, "Alexander, we're at Dinosaur Ridge, and they're having a big celebration."

"What's going on?" Alexander asked.

"Look at the banner above the entrance. It says, OLDEST DINOSAUR BABIES RECENTLY DISCOVERED. Let's go inside and find out what is going on!" his mother exclaimed.

Alexander and his mother went inside the museum. The museum had many interactive displays for the children. At the first stop, children got to practice digging up their own dinosaur eggs, like paleontologists do in the field. The display was filled with giant sandboxes and paintbrushes for the children to use in the sand, where they found their own dinosaur fossils.

"Look, Mommy," Alexander exclaimed, "I'm a paleontologist!" Alexander's mom smiled at him as he brushed away the sand, proudly displaying his discovery.

While the children uncovered their own fossilized dinosaur eggs, they also listened to a guide from the museum tell them about the recent scientific discovery advertised on the banner.

"The earliest dinosaurs were alive about 230 million years ago. A paleontologist named Dr. Robert Reisz and his team found the oldest fossilized dinosaur embryos ever discovered," the guide said. "*Fossilized* means 'turned to rock or stone,' and *embryos* are 'baby dinosaurs before they hatch from their eggs.' So you could say that baby dinosaurs still in their eggs, were discovered.

"These fossilized dinosaur embryos are from more than 190 million years ago. This is the oldest discovery of baby dinosaur eggs ever! In April 2013 the scientist reported on his discovery from China. Dr. Reisz believes the fossils are probably from a dinosaur called *Lufengosaurus*, pronounced Loo-FUNG-oh-SAWR-us. It means 'Lufeng lizard.' Lufeng is a part of China. These were large dinosaurs that only ate plants. They grew big: six meters in length, which is about half the size of a school bus.

"Dr. Reisz's research showed that when the baby dinosaurs were growing inside the eggs, the babies were kicking their legs, like baby humans and other mammals and birds do.

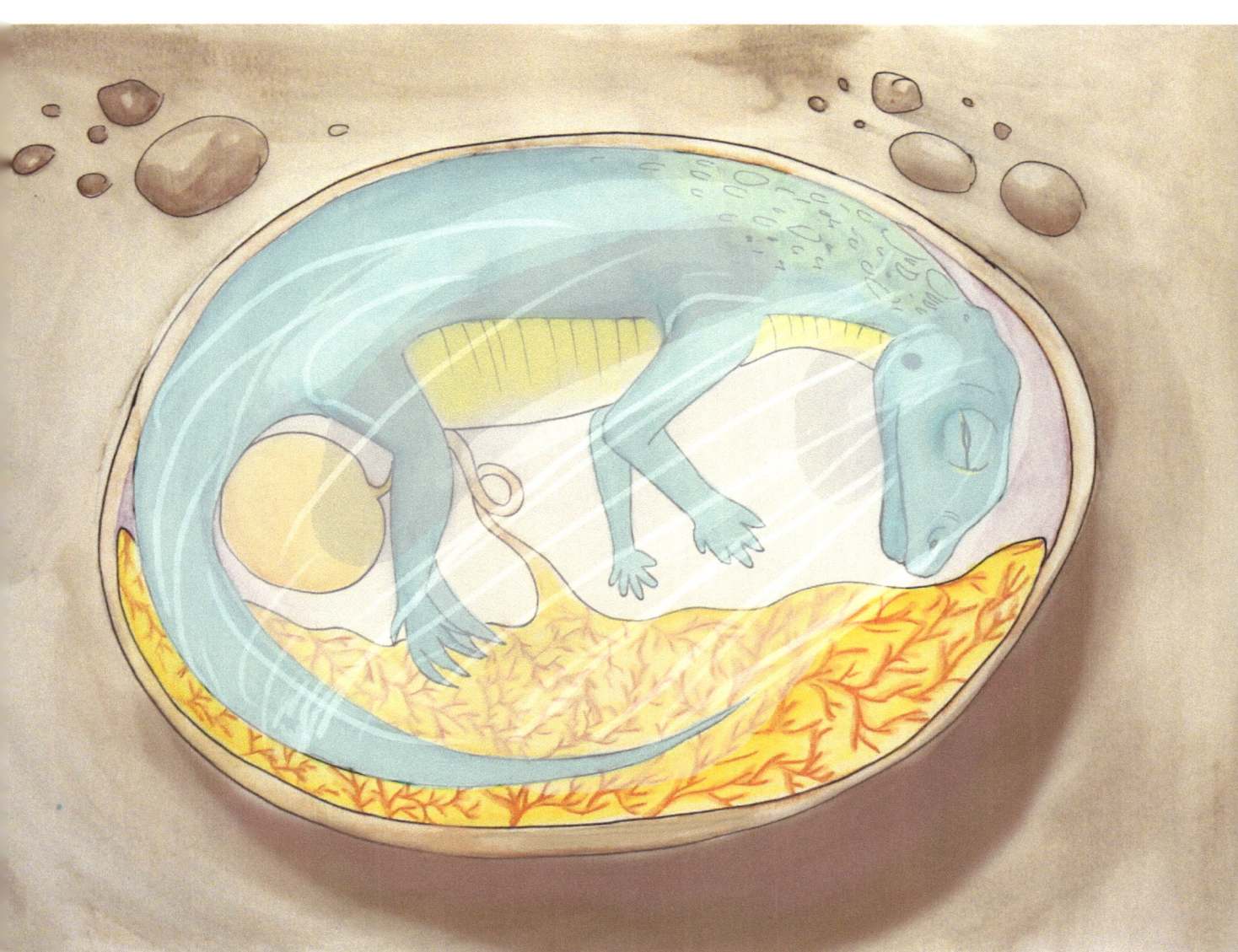

This is probably so the baby dinosaurs could move once they hatched.

These baby dinosaurs grew really fast while in their eggs—faster than any birds or mammals living today. The dinosaur's mother laid her eggs in a nest on the ground beside the nests of other dinosaur mothers, usually near water. Inside the egg was everything the baby dinosaur needed to grow."

The children all enjoyed listening to the guide. Alexander began to explore the rest of the museum. A life-sized replica of a female Lufengosaurus was the next display. The dinosaur waved her tail in the air above the children, who jumped up and down, trying to catch it. Of course, none of the children could grasp the tail because it was so high. But they all had a lot of fun attempting to catch it as it swung above their heads.

At another display, a giant head of the dinosaur was placed on the ground, and her mouth was wide open. It was so wide that Alexander could put his head into her mouth.
"Look, Mommy," said Alexander. "The Lufengosaurus is going to eat me!"
Alexander's mother laughed and took a picture.

They moved to the last display in the museum: a giant mural that all of the children could color. Alexander picked his favorite color from the crayons—magenta sky—and began to color. He laughed and enjoyed himself with the other children. His mother watched and talked with the other parents.

Alexander and his mother left the museum and got back into their car.

"Alexander, that was an interesting visit to Dinosaur Ridge. Did you like it?" asked his mom.

"Yes," said Alexander. "I got to see the oldest dinosaur eggs ever discovered!"

They started their drive home. Alexander's mother put him into the backseat of the car. Because it was afternoon and there was lots of sunshine, the car was hot. Alexander pulled his shirt halfway up so that he could cool off. He looked down at his belly button.

"Mommy, do dinosaurs have belly buttons?" Alexander asked.

Alexander's mother glanced in the rearview mirror. She had to think about the answer to his question.

She answered, "Well, Alexander, every person has a belly button. Boys and girls do. Moms and dads do."

Alexander quickly asked his mother, "But what if I had a sister? Would she have a belly button?"

His mom said, "Yes, Alexander. Even she would have a belly button. Alexander, let's talk about how you got your belly button, and we will see if that answers your question.

"Before you were born, you grew inside me—inside a part of my body that only women have, called a womb or uterus." Alexander asked, "You mean I don't have a womb, Mommy?" His mother answered, "That's right, Alexander. Boys and men do not. While you were growing in my womb, I was pregnant.

"A pregnancy is for about nine months, or around forty weeks. You started out very small—the size of a grain of sand, or much smaller than a snowflake—and you grew into a baby."

"But Mommy," said Alexander, "how can something the size of a snowflake grow into a baby? That doesn't seem possible."

His mother answered, "Alexander, it's a miracle. What helps a baby to grow inside the womb are a placenta, an umbilical cord, and amniotic fluid."

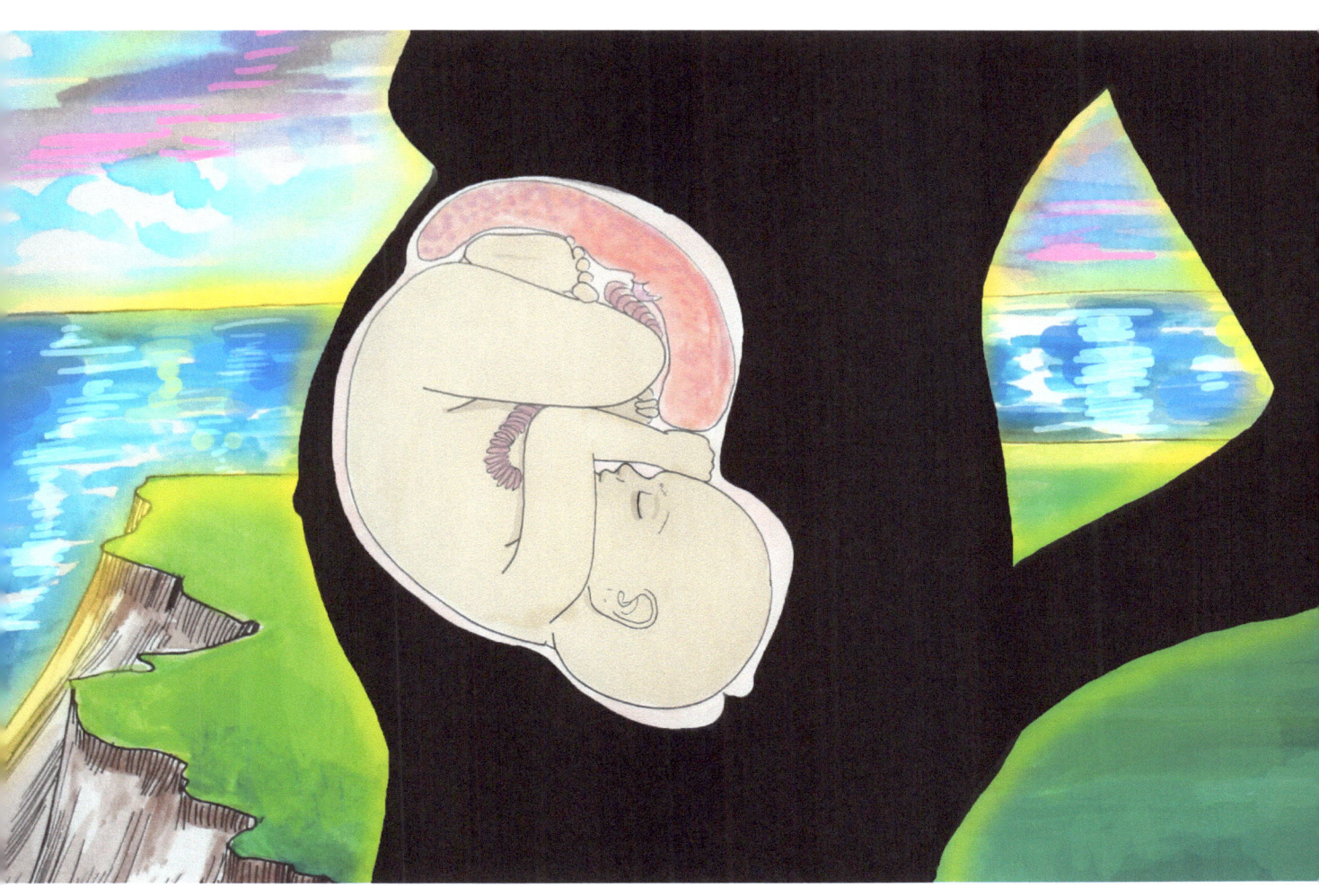

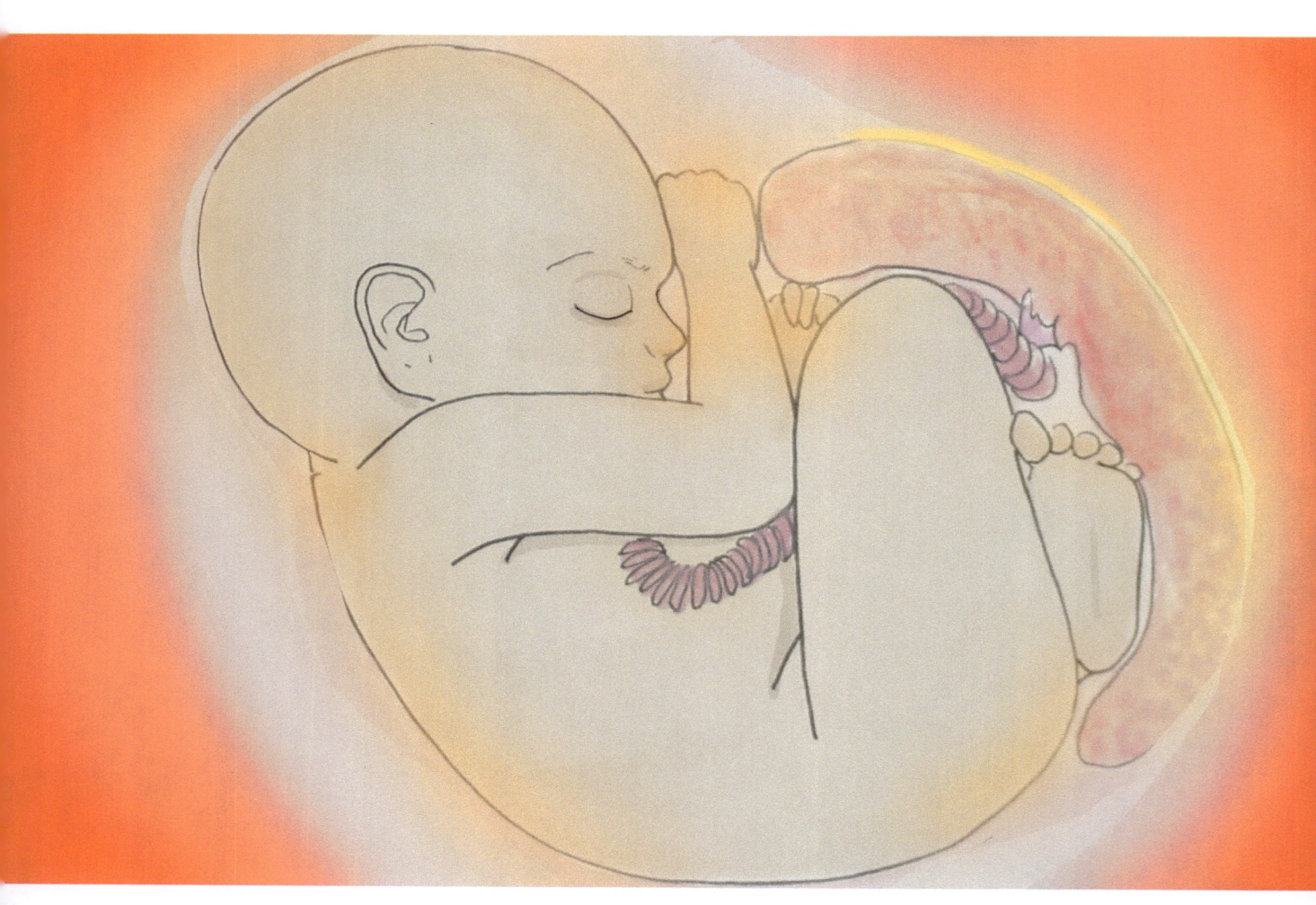

"It sounds crowded in there, Mommy!" Alexander said, making a funny face.

His mother smiled at him, glancing in the rearview mirror again. "It was," she said, "but it was all necessary for you to be able to eat, breathe, and grow. The womb is like a house for the baby. In the womb is everything the baby needs. The baby gets its food from the umbilical cord, which is attached at the baby's navel.

Do you understand, Alexander?"

"Well, did you like having me inside of you, Mommy?"
Alexander asked.

"Yes, Alexander. I loved being pregnant with you. It was amazing!" said his mother.

"But Mommy, what about my belly button? Don't forget about that part," said Alexander.

His mother answered, "When a baby is born, it's born with its umbilical cord.

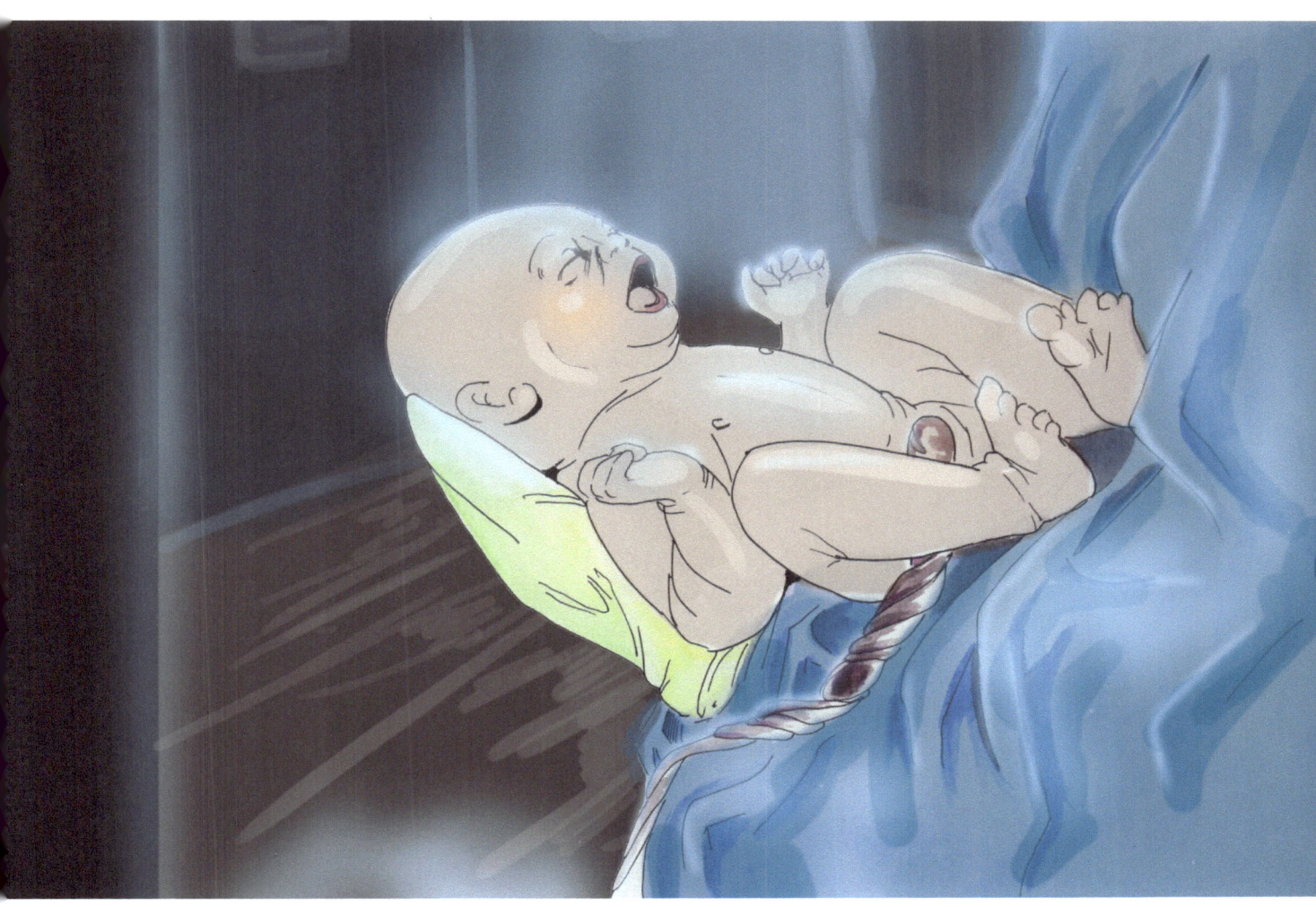

"Attached to that is a placenta, which is also called an afterbirth. After the baby is born, the umbilical cord is cut. At the baby's birth, about an inch of the umbilical cord is left with the baby for a few weeks. Then it falls off on its own, and the baby is left with his or her own belly button!

"So that is how you got your belly button, Alexander. First I was pregnant with you, and you grew in my uterus. Then you were born, and your umbilical cord was cut. And after a short amount of time, it fell off, and you were left with your belly button."

"Alexander, do you think you know the answer to your question now?" his mother asked.

"Yes I do, Mommy. I have a belly button, but dinosaurs—because they hatched from eggs and did not grow inside their mommy—do not have belly buttons!"

"That is right!" his mother said.

"I had fun today, Mommy," said Alexander.

"I did too, and we are almost home," said his mother. "Alexander, always remember as long as you live, that having you as my son is the best part of my life. I love you, Alexander."

"I love you too, Mommy," said Alexander.